I0489832

Deep Space Radiotrophs – Extraterrestrial Alien Life

Expanded 2nd Edition

Author: Michael F. Zahra

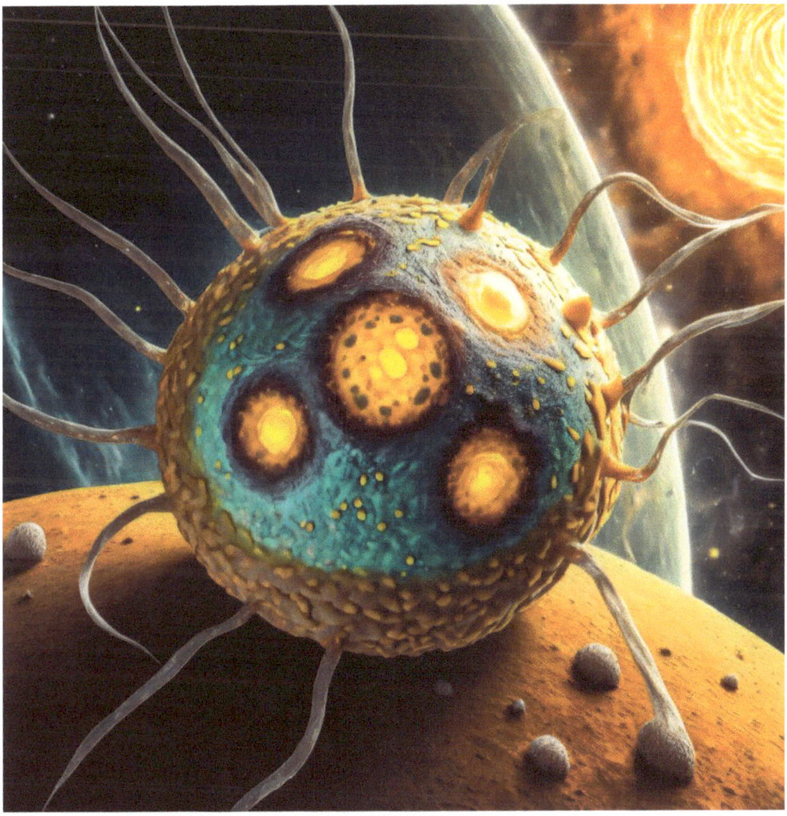

All Radiotrophic Images Contained Herein are Artistic Impressions Only.

This expanded 2nd Edition adds further information on exoplanets, moons and comets in the universe as well as non-water-based life alternative chemistry.

This book is dedicated to the many people smarter than me with whom I've had the privilege to work. You know who you are.

Contents

Astrobiology and Extraterrestrial Life   4
Introduction to Radiotrophs   13
Basic Physiology   16
Radiation Absorption Mechanism   16
Metabolic Processes   18
Radiation Protection Mechanisms   28
Nutrient Requirements   32
Reproduction and Growth   33
Ecological Role   36
Evolutionary History   40
Earth-based Analogs   45
Biotechnological Applications   52
Non-Water-Based Life Chemistry   55
Further Research   79
Cautions, Considerations and Biosafety   82
Conclusion   85
References   87

Deep Space Radiotrophs –
Extraterrestrial Alien Life

Expanded 2nd Edition

Author: Michael F. Zahra

This book proposes a hypothetical, single-celled (or potentially complex multicellular) organism with a unique biochemistry adapted to utilize radiation as its primary energy source for metabolic processes.

Astrobiology and Extraterrestrial Life

Astrobiology, the interdisciplinary study of the origin, evolution, distribution, and future of life in the universe, encompasses a wide range of scientific disciplines aimed at understanding the potential for life beyond Earth. Central to astrobiology is the exploration of habitable environments in our solar system and beyond, the search for biosignatures indicating past or present life, and the development of theoretical frameworks for understanding the diversity and adaptability of life forms in extreme environments.

Habitability of Extraterrestrial Environments: The search for extraterrestrial life begins with identifying environments that could potentially support life, as we know it, or otherwise. Habitable environments may include planets, moons, comets or exoplanets within the Habitable Zone$^\alpha$ of their host stars, where conditions allow for the presence of liquid water - a key prerequisite for life, as we understand it. Additionally, astrobiologists explore other potential habitats, such as subsurface oceans, hydrothermal vents, icy moons, and regions with geothermal activity, where alternative forms of life might thrive. Non-water-based environments will also be discussed in detail.

Potential Habitats for Radiotrophic Organisms: In the context of Radiotrophic organisms, astrobiologists consider environments with high levels of ionizing radiation as potential habitats for extremophiles capable of harnessing radiation energy for metabolism. These environments may include regions of intense cosmic radiation in space, such as cosmic ray-dominated regions in interstellar space or near pulsars and supernovae. On planetary bodies, potential habitats for Radiotrophic organisms could include subsurface environments with radioactive isotopes, regions near radioactive sources, or locations with high natural background radiation.

Search for Biosignatures: Astrobiologists employ a variety of techniques to search for biosignatures - indicators of past or present life - in extraterrestrial environments. Biosignatures may include chemical signatures associated with biological activity, such as organic molecules, isotopic ratios, or patterns of molecular complexity. Additionally, astrobiologists investigate the potential for remote detection of biosignatures through telescopic observations, spectroscopic analysis, and the search for biomarkers in planetary atmospheres or surface materials.

Experimental Approaches and Analogs: In addition to observational studies, astrobiologists conduct laboratory experiments and field studies to simulate extraterrestrial environments and investigate the survivability and adaptability of life forms under extreme conditions. Analog environments on Earth, such as Mars analog sites, deep-sea hydrothermal vents, or radiation-contaminated environments, serve as testbeds for studying the potential habitability of extraterrestrial environments and the adaptations of extremophiles, including Radiotrophic organisms.

Future Missions and Exploration: Future space missions and exploration efforts, including robotic missions to Mars, missions to icy moons like Europa and Enceladus, and the search for exoplanets with habitable conditions, play a crucial role in advancing astrobiology research and the search for extraterrestrial life. These missions aim to characterize planetary environments, assess their habitability, and search for biosignatures that could provide clues to the presence of life beyond Earth.

Astrobiology represents a multidisciplinary field at the intersection of astronomy, biology, and planetary science, aimed at understanding the potential for life beyond Earth. The search for extraterrestrial life encompasses the exploration of habitable environments, based on current or new definitions, the search for biosignatures, experimental studies of extremophiles, and future space missions and exploration efforts. In the context of Radiotrophic organisms, astrobiologists investigate the potential habitability of environments with high levels of ionizing radiation and the adaptations of extremophiles capable of harnessing radiation energy for metabolism.

By integrating insights from diverse scientific disciplines, astrobiology offers a comprehensive framework for exploring the origins, evolution, and diversity of life in the universe, including the potential existence of Radiotrophic organisms in extraterrestrial environments.

The Drake equation is a probabilistic argument used to estimate the number of *active, communicative* (i.e. intelligent and multicellular) extraterrestrial civilizations in the Milky Way galaxy:

$$N = R_* \cdot f_p \cdot n_e \cdot f_l \cdot f_i \cdot f_c \cdot L$$

Where,

N = the number of civilizations in the Milky Way galaxy with which communication might be possible;

and

R_* = the average rate of star formation in our galaxy.
f_p = the fraction of those stars that have planets.
n_e = the average number of planets that can potentially support life per star that has planets.
f_l = the fraction of planets that could support life that actually develop life at some point.

f_i = the fraction of planets with life that go on to develop intelligent life (civilizations).

f_c = the fraction of civilizations that develop a technology that releases detectable signs of their existence into space.

L = the length of time for which such civilizations release detectable signals into space.

The equation was formulated in 1961 by Frank Drake, not for purposes of quantifying an absolute number of civilizations, but as a way to stimulate dialogue at the first scientific meeting on the Search for Extraterrestrial Intelligence (SETI). It is more properly thought of as an approximation than as a serious attempt to determine a precise number.

Criticism of the Drake equation focuses on the fact that the estimated values for several of its factors are highly conjectural, the combined multiplicative effect being that the uncertainty associated with any derived value is so large that the equation cannot be used to draw firm, absolute conclusions.

Drake's proposed estimate is shown below, but numbers on the right side of the equation are agreed as speculative and open to substitution:

$$10{,}000 = 5 \cdot 0.5 \cdot 2 \cdot 1 \cdot 0.2 \cdot 1 \cdot 10{,}000$$

It is presented herein as more of a novelty to stimulate the reader's contemplation of extraterrestrial life in our local Milky Way galaxy.

Drake looked at *intelligent, communicative* life. Proposed Radiotrophic life would likely be single-celled and potentially be more abundant, not requiring a host planet/moon/comet specifically in the Habitable Zone.

The number of superclusters in the visible universe[β] is ~10 million. The number of galaxy groups in the visible universe ~25 billion. The number of large galaxies in the visible universe ~350 billion. The number of dwarf galaxies in the visible universe ~7 trillion.

While some stars have no planets, others, like our Sun, have multiple planets. Common assumptions average ~1 planet per star. The number of stars (and therefore planets), in only the visible universe is estimated to be 3×10^{22}, or 30,000,000,000,000,000,000,000.

It is likely that the galaxies within the observable universe represent only a small fraction of the galaxies in the inflationary universe (observed plus unobserved). According to the theory of cosmic inflation, if it is assumed that inflation began about 10^{-37} seconds after the Big Bang and that the pre-inflation size of the universe was approximately equal to the speed of light times its age, that would suggest that at present, the inflationary universe's size is at least 1.5×10^{34} light-years, or at least 3×10^{23} times the radius of the observable universe.

This could mean the estimated number of stars (and possibly planets) in the inflationary universe is 10^{100}.

In 2013, astronomers reported, based on Kepler data, that there could be as many as 40 billion Earth-sized planets orbiting in the habitable zones of Sun-like stars and red dwarfs in just the Milky Way alone. Some are "close", with several within 30 light-years of our Sun.

Proxima Centauri b, located only about 4.2 light-years from Earth in the constellation of Centaurus, is the nearest known exoplanet, and is orbiting in the habitable zone of its red dwarf star.

Many stars have been confirmed to possess one or even multiple HZ planets. The HZ is also of particular interest to the emerging field of habitability of natural satellites, because planetary-mass moons in the HZ likely significantly outnumber planets.

Additionally, while the are only 3,910 known comets, according to the European Space Agency, the number of comets in the entire universe is estimated to be one trillion or 10^{12}.

Given the massive size of the entire inflationary universe, including planets, their moons and comets, the mathematical case for life elsewhere becomes quite compelling.

Radiotrophic life is likely simple, single-celled, and therefore would not have the ability to travel, communicate or make contact with Earth. This bodes well for its existence, but being yet undetected.

The Fermi Paradox, named after physicist Enrico Fermi, is the discrepancy between the lack of conclusive evidence, such as radio signals or other forms of communication, of advanced extraterrestrial life and the apparently high likelihood of its existence. It is a conflict between the argument that scale and probability seem to favour intelligent life being somewhat common in the universe, and the total lack of firm evidence of intelligent life having ever arisen anywhere other than on Earth.

In the context of Radiotrophs, the Fermi Paradox could be considered from several perspectives:

Uniqueness of Radiotrophs: If Radiotrophs are indeed a possible form of life that can exist in deep space, their existence could potentially provide a unique perspective on the Fermi Paradox. Their ability to thrive in environments with high levels of ionizing radiation, such as near black holes or in highly irradiated regions of space, could lead to the emergence of complex ecosystems that differ significantly from those found on planets like Earth.

Potential for Detection: Radiotrophs might emit detectable signatures of their presence through their interactions with radiation or byproducts of their metabolic processes. For example, they could emit characteristic radiation signatures that could be detected by telescopes or other instruments. If Radiotrophs are widespread in the universe, their potential detectability could influence our understanding of the Fermi Paradox by providing new avenues for searching for extraterrestrial life.

Constraints on Technological Civilizations: The existence of Radiotrophs could contribute to potential solutions to the Fermi Paradox by introducing additional factors that may constrain the development of technological civilizations. For example, if Radiotrophs dominate certain regions of space, they could compete with or impede the development of other forms of life, including future intelligent civilizations capable of technological advancement.

Radiotrophs could offer new insights into the Fermi Paradox by expanding our understanding of the diversity of life in the universe and its implications for the search for extraterrestrial intelligence.

Footnotes to this section:
[α] The Habitable Zone (HZ), or more precisely the Circumstellar Habitable Zone (CHZ), in astronomy and astrobiology, is defined as the range of orbits around a star within which a planetary surface can support liquid water given sufficient atmospheric pressure.

[β] The visible universe is defined as a 46.5 billion light-year, sphere-shaped subset region of the inflationary universe comprising all matter that can be observed from Earth, or space-based telescopes and exploratory probes, at the present time. Electromagnetic radiation from these objects has had time to reach Earth since the beginning of the Big Bang.

Introduction to Radiotrophs

It's theoretically feasible. There are extremophiles on Earth that can survive in extreme conditions, including radiation-rich environments. It's not implausible to imagine life forms in the vast universe adapting to use radiation as an energy source.

Hypothetically, such life forms might have adaptations to withstand high levels of radiation, possibly having thick, radiation-resistant outer layers or specialized internal structures to protect vital structures or organs. They might be highly efficient at repairing DNA damage caused by radiation or have mechanisms to harness radiation directly for metabolic processes. Their appearance could vary widely depending on their evolutionary path and the specific environment in which they evolved.

One hypothetical mechanism could involve specialized molecules that can absorb and convert radiation into chemical energy usable by the organism. These molecules could act as "radiosynthetic pigments" similar to photosynthetic pigments in plants, but instead of converting light energy into chemical energy, they would convert radiation energy. Another possibility could involve symbiotic relationships with radiation-absorbing microorganisms or even incorporating radiation-absorbing minerals into their cellular structure.

They could be single-celled organisms, multicellular plants, animals, or even more complex organisms similar to Earth's mammals, fish, or reptiles. The form of these hypothetical life forms would depend on various factors including the environment they inhabit, the availability of radiation as an energy source, and the evolutionary pressures shaping their development. For example, in environments with abundant radiation and few resources, single-celled organisms or simple multicellular organisms might be more common, whereas in more complex ecosystems with diverse energy sources, more complex organisms could evolve to utilize radiation as an additional energy source.

It's possible for organisms to survive based solely on using radiation for metabolic processes, similar to how plants use photosynthesis. However, the specific requirements for such organisms would depend on the details of their biology and the environment they inhabit. While plants also require water and nutrients from soil, organisms relying on radiation as their primary energy source might have evolved alternative mechanisms for obtaining essential nutrients or might inhabit environments where those nutrients are available through other means, such as through interactions with other organisms or geological processes.

In this speculative scenario, we'll explore hypothetical life forms existing in an environment with high levels of ionizing radiation, such as near a highly active stellar object or within a region of intense cosmic radiation.

Cosmic radiation in deep space is composed of various types of ionizing radiation, including protons, alpha particles, beta particles (electrons and positrons), gamma rays, and cosmic rays (high-energy charged particles such as protons, helium nuclei, and heavier ions). These particles originate from a variety of sources, including the sun, other stars, and even distant astrophysical phenomena such as supernovae and active galactic nuclei.

The composition and energy spectrum of cosmic radiation can vary depending on factors such as the source of the radiation, the distance from the source, and the interactions of the radiation with interstellar matter. For example, near Earth, the sun is a significant source of cosmic radiation, which includes protons, alpha particles, and a small fraction of heavier ions. In deep space, away from the influence of the solar wind and magnetic fields, cosmic radiation consists of a more diverse mix of high-energy particles, including cosmic rays of various energies and origins.

Basic Physiology
Our hypothetical life form, let's call it "Radiotroph," is a single-celled organism with a unique biochemistry adapted to utilize radiation as its primary energy source for metabolic processes. Radiotrophs are microorganisms with a spherical shape, similar in size to certain types of bacteria found on Earth.

Radiation Absorption Mechanism
Radiotrophs, or organisms capable of utilizing ionizing radiation as an energy source, possess specialized cellular structures and biochemical mechanisms that enable them to absorb and convert ionizing radiation directly into chemical energy for metabolic processes. Understanding the radiation absorption mechanisms employed by Radiotrophs is essential for studying their unique biology and metabolic pathways.

Specialized Cell Membrane Proteins: Radiotrophs exhibit specialized cell membrane proteins that play a crucial role in the absorption of ionizing radiation from their environment. These proteins may act as radiation receptors or ion channels, facilitating the entry of ionizing radiation into the cell interior. These proteins are likely to possess unique structural features that enable them to interact with ionizing radiation and initiate downstream biochemical processes.

Intracellular Structures: Within the intracellular environment, Radiotrophs have specialized structures or organelles dedicated to the absorption and utilization of ionizing radiation. These structures may include radiation-absorbing pigments, electron transport chains, or enzyme complexes that participate in radiation capture and energy conversion processes. These intracellular structures are optimized to maximize the efficiency of radiation absorption and harness the energy released for metabolic activities.

Analogous to Chlorophyll Molecules: The radiation absorption mechanisms in Radiotrophs may bear resemblance to the light-capturing mechanisms employed by chlorophyll molecules in photosynthetic organisms. Chlorophyll molecules in plants and algae absorb photons of light energy, initiating photosynthesis and converting light energy into chemical energy in the form of adenosine triphosphate (ATP) and nicotinamide adenine dinucleotide phosphate (NADPH). Similarly, Radiotrophs possess specialized molecules or structures that capture ionizing radiation (such as gamma rays or cosmic rays) and convert it into chemical energy through analogous biochemical processes.

Conversion of Ionizing Radiation to Chemical Energy: Upon absorption of ionizing radiation, Radiotrophs undergo biochemical reactions that convert the energy of ionizing radiation into chemical energy stored in high-energy molecules such as ATP or reduced coenzymes (e.g., nicotinamide adenine dinucleotide (NADH) or NADPH). These biochemical reactions involve electron transfer processes, generation of reactive oxygen species (ROS), and activation of metabolic pathways that utilize the energy released by ionizing radiation to drive cellular metabolism and growth.

Adaptations for Radiation Tolerance: Radiotrophs have evolved specialized adaptations to cope with the damaging effects of ionizing radiation on cellular components, including DNA, proteins, and lipids. These adaptations may include mechanisms for DNA repair, scavenging of free radicals, activation of antioxidant defenses, and regulation of metabolic pathways to minimize radiation-induced damage and maintain cellular viability in high-radiation environments.

Radiotrophs possess specialized cell membrane proteins and intracellular structures that enable them to absorb and convert ionizing radiation directly into chemical energy for metabolic processes. These radiation absorption mechanisms are analogous to the light-capturing mechanisms employed by chlorophyll molecules in photosynthetic organisms but adapted for capturing and utilizing ionizing radiation as an energy source. Understanding the radiation absorption mechanisms in Radiotrophs is essential for elucidating their unique biology, metabolic pathways, and adaptations to high-radiation environments, with implications for astrobiology, biotechnology, and the search for life beyond Earth.

Metabolic Processes
Radiotrophs have evolved metabolic pathways that utilize the absorbed radiation to drive essential cellular

processes, including growth, reproduction, and maintenance. These metabolic pathways involve unique enzymes and biochemical reactions specifically adapted to harness the energy from ionizing radiation.

Radiation Absorption Enzymes: These enzymes are embedded within specialized cell membrane proteins and intracellular structures designed to absorb ionizing radiation. They could be composed of unique protein subunits with metal cofactors that facilitate the capture and transfer of radiation energy to downstream metabolic pathways.

Radiation Conversion Enzymes: Once absorbed, ionizing radiation is converted into chemical energy through a series of enzymatic reactions. These enzymes could catalyze the conversion of radiation energy into high-energy chemical bonds, such as ATP, through processes analogous to phosphorylation or redox reactions.

Radiation-Driven Metabolism: Radiotrophs utilize the energy derived from ionizing radiation to drive essential metabolic processes. This includes the synthesis of ATP for cellular energy, as well as the production of precursor molecules for biosynthesis, such as amino acids, nucleotides, and lipids.

Radiation-Activated Pathways: Specific metabolic pathways are activated in response to radiation exposure, allowing Radiotrophs to efficiently utilize the energy from ionizing radiation. These pathways may involve unique enzyme cascades and regulatory mechanisms tailored to the radiation-rich environment.

Radiation-Induced DNA Repair Enzymes: To mitigate the damage caused by ionizing radiation, Radiotrophs possess specialized enzymes involved in DNA repair. These enzymes rapidly detect and repair radiation-induced DNA lesions, such as single-strand breaks, double-strand breaks, and base modifications, to maintain genomic integrity and ensure cell viability.

Radiation-Responsive Antioxidant Enzymes: Radiotrophs have evolved antioxidant defense systems to counteract the harmful effects of ROS generated by ionizing radiation. These systems include enzymes such as superoxide dismutase, catalase, and peroxidases, which scavenge ROS and protect cellular components from oxidative damage.

Radiation-Resistant Proteins: Radiotrophs produce structural proteins that confer resilience to ionizing radiation. These proteins may contain unique amino acid compositions or post-translational modifications that enhance their stability and minimize radiation-induced damage to cellular structures, including membranes, organelles, and cytoskeletal elements.

Protein Subunits: Within the specialized cell membrane proteins and intracellular structures of Radiotrophs, there exist protein subunits that play a crucial role in the absorption and conversion of ionizing radiation into chemical energy. These protein subunits are composed of long chains of amino acids arranged in specific sequences and configurations to facilitate their function.

Metal Cofactors: Metal cofactors are essential components of these protein subunits, contributing to their structural stability and catalytic activity in radiation absorption and conversion processes. These metal ions serve as integral components of the protein structure, often coordinated by specific amino acid residues within the protein's active site.

In deep space, the types of metals that could potentially serve as cofactors in the protein subunits of Radiotrophs would depend on several factors, including their abundance in the interstellar medium, their suitability for coordination chemistry within biological systems, and their ability to facilitate radiation absorption and conversion processes. While specific knowledge of metal availability in deep space is limited, we can speculate on some potential candidates based on known astrophysical processes and observations.

Cadmium (Cd): Cadmium is a toxic heavy metal, but in trace amounts, it could serve as a cofactor in specialized enzymes adapted to high-radiation environments, potentially playing a role in detoxification or energy metabolism pathways in Radiotrophic organisms.

Cobalt (Co): Cobalt is relatively rare in the universe compared to iron and nickel but is still present in interstellar gas and dust. It is known to participate in the active sites of certain metalloenzymes involved in energy metabolism and radical reactions. Cobalt is known to form stable complexes with organic molecules.

Copper (Cu): Copper is less abundant in the interstellar medium but is still present in stellar atmospheres and supernova remnants. It is known to participate in redox reactions and electron transfer processes in biological systems, making it a potential candidate for radiation-converting enzymes.

Iron (Fe): Iron is one of the most abundant metals in the universe and is commonly found in interstellar clouds, supernova remnants, and cosmic dust grains. It exhibits versatile coordination chemistry and is known to serve as a cofactor in many biological metalloenzymes on Earth, like human blood.

Magnesium (Mg): Magnesium is essential for the structure and function of nucleic acids, and it could serve as a cofactor in enzymes involved in DNA repair and replication in Radiotrophic organisms.

Manganese (Mn): Manganese is also less abundant than iron and nickel but is still found in interstellar environments, especially in association with oxygen-rich dust grains. It plays essential roles in oxygen-evolving complexes of photosynthetic organisms and could potentially serve as a cofactor in radiation-absorbing enzymes.

Molybdenum (Mo): Molybdenum is an essential trace element that can act as a cofactor in enzymes involved in redox reactions and nitrogen metabolism, potentially playing a role in Radiotrophic energy conversion processes.

Nickel (Ni): Nickel is another abundant metal in the universe, commonly found in meteorites, interstellar dust grains, and stellar atmospheres.

It forms stable complexes with organic ligands and has been proposed as a potential cofactor in certain biological processes.

Selenium (Se): Selenium is a trace element with antioxidant properties and could potentially act as a cofactor in enzymes involved in detoxification of reactive oxygen species generated by ionizing radiation in Radiotrophic organisms.

Tungsten (W): Tungsten is known for its role in certain bacterial enzymes involved in anaerobic metabolism, and it could hypothetically serve as a cofactor in specialized enzymes adapted to radiation-rich environments in Radiotrophic organisms.

Vanadium (V): Vanadium is a transition metal with redox-active properties and could potentially participate in electron transfer reactions and energy metabolism pathways in Radiotrophic organisms exposed to high levels of ionizing radiation.

Zinc (Zn): Zinc is a versatile metal ion that can participate in a variety of enzymatic reactions, including catalysis and structural stabilization of proteins involved in radiation absorption and metabolism. At least 1000 human proteins (out of ~20,000) contain zinc-binding protein domains.

Zirconium (Zr): Zirconium is a transition metal that forms stable complexes with organic ligands, and it could potentially serve as a cofactor in enzymes involved in radiation absorption and energy conversion pathways in Radiotrophic organisms.

These metal cofactors illustrate the diversity of metals that could potentially contribute to the biochemistry and metabolism of Radiotrophic organisms in radiation-rich environments. Further research is needed to explore their specific roles and biochemical functions in Radiotrophic biology.

Coordination Chemistry: The metal cofactors within the protein subunits of Radiotrophs exhibit complex coordination chemistry, characterized by precise geometric arrangements of amino acid residues and metal ions. This coordination enables efficient capture and transfer of radiation energy to downstream biochemical pathways, while also providing structural integrity to the protein complex.

Metalloenzymes: These protein subunits, known as metalloenzymes, contain metal cofactors that are essential for their catalytic function in radiation-driven metabolic processes. Examples of metalloenzymes found in Radiotrophs include radiation-absorbing enzymes involved in initial radiation capture, as well as radiation-converting enzymes responsible for subsequent energy conversion reactions.

Metal Ion Specificity: The selection of specific metal ions as cofactors within these protein subunits is critical for their function and efficiency in radiation absorption and conversion. Different metal ions exhibit unique properties, such as redox potential, coordination geometry, and ligand binding affinity, which influence the overall performance of the metalloenzymes in radiation-driven metabolic pathways.

Structural Adaptations: The protein subunits containing metal cofactors undergo structural adaptations to accommodate the presence of metal ions and optimize their function in radiation absorption and conversion. These structural adaptations may involve conformational changes, protein-protein interactions, and lipid-protein interactions within the cell membrane and intracellular compartments of Radiotrophs.

The protein subunits with metal cofactors found in Radiotrophs represent sophisticated molecular machinery adapted to harness ionizing radiation for metabolic processes. Through complex coordination chemistry and structural adaptations, these metalloenzymes play a central role in capturing and converting radiation energy into chemical energy, enabling the survival and proliferation of Radiotrophs in radiation-rich environments.

Radiation Protection Mechanisms

To prevent excessive damage from radiation, Radiotrophs have evolved robust DNA repair mechanisms and antioxidant systems that can efficiently repair DNA damage caused by ionizing radiation. Additionally, their cell membranes are reinforced with radiation-resistant materials to provide structural protection.

These mechanisms involve a combination of structural, biochemical, and molecular adaptations aimed at mitigating radiation-induced damage to cellular components.

DNA Repair Systems: One of the primary radiation protection mechanisms involves DNA repair systems that detect and repair damage to the genetic material caused by ionizing radiation. These systems include various repair pathways such as base excision repair, nucleotide excision repair, and double-strand break repair. Specialized enzymes recognize and remove damaged DNA bases or facilitate the rejoining of broken DNA strands, restoring genomic integrity and preventing mutations.

Antioxidant Defenses: Ionizing radiation can generate ROS within cells, leading to oxidative stress and damage to cellular structures. To counteract this, organisms have evolved antioxidant defense systems that scavenge ROS and neutralize their harmful effects. Enzymes such as superoxide dismutase, catalase, and peroxidases catalyze the conversion of ROS into less reactive species, protecting cellular components from oxidative damage.

Melanin Pigmentation: Melanin, a dark pigment found in certain organisms, serves as a natural sunscreen that absorbs and dissipates ionizing radiation. Melanin pigmentation provides structural protection to cellular components by shielding them from direct exposure to radiation and reducing the penetration of harmful radiation into tissues. This protective mechanism is particularly relevant in organisms inhabiting environments with high levels of ultraviolet (UV) and ionizing radiation.

Cellular Repair Machinery: Beyond DNA repair systems, cells possess a range of repair and maintenance mechanisms to counteract radiation-induced damage to proteins, lipids, and other cellular components. Chaperone proteins assist in protein folding and refolding, helping to repair misfolded or denatured proteins caused by radiation. Lipid repair enzymes mitigate damage to cell membranes, ensuring their structural integrity and function in maintaining cellular homeostasis.

Radiation-Resistant Structures: Some organisms have developed radiation-resistant structures that provide physical protection against ionizing radiation. These structures may include thick cell walls, specialized outer layers, or extracellular matrices that act as barriers to radiation penetration. Additionally, organisms inhabiting radiation-rich environments may exhibit unique morphological adaptations, such as reduced cellular size or altered cellular organization, to minimize radiation exposure to vital cellular structures.

Efflux Mechanisms: Certain organisms possess efflux mechanisms that actively remove toxic substances, including radiation-induced damage products, from their cells. Efflux pumps and transporters facilitate the expulsion of harmful compounds, maintaining cellular homeostasis and preventing the accumulation of toxic metabolites that can interfere with cellular function.

These radiation protection mechanisms collectively contribute to the survival and adaptation of organisms in radiation-rich environments, including hypothetical

Radiotrophic organisms, by minimizing the deleterious effects of ionizing radiation on cellular structures and functions.

Nutrient Requirements

Since Radiotrophs primarily rely on radiation for energy, their nutrient requirements are minimal compared to traditional, known carbon-based life forms. However, they may still require trace elements and minerals for cellular functions, which they obtain from their environment through passive diffusion or specialized uptake mechanisms.

Reproduction and Growth
Radiotrophs reproduce asexually through binary fission, splitting into two identical daughter cells. Under optimal conditions with abundant radiation, they can reproduce rapidly, forming colonies of interconnected cells. Growth rates may vary depending on the availability of radiation and other environmental factors.

Asexual Reproduction: In radiation-rich environments, where the survival of individual organisms may be challenging due to the potential for radiation-induced damage, Radiotrophic organisms may predominantly employ asexual reproduction. Asexual reproduction methods such as binary fission, budding, or fragmentation allow for the rapid proliferation of genetically identical offspring without the need for mating or genetic recombination. This mode of reproduction ensures the efficient transmission of beneficial adaptations to subsequent generations, facilitating population growth and adaptation to radiation-rich environments.

Rapid Reproductive Cycles: Radiotrophic organisms may exhibit adaptations that enable them to undergo rapid reproductive cycles in response to fluctuations in radiation levels or environmental conditions. Shortened generation times, accelerated growth rates, and enhanced reproductive output allow for the rapid colonization and exploitation of radiation-rich habitats. These adaptations ensure the continued survival and expansion of Radiotrophic populations in dynamic and challenging environments.

Reproductive Resilience: Radiotrophic organisms may possess mechanisms to maintain reproductive resilience in the face of radiation-induced damage to reproductive tissues or germ cells. Enhanced DNA repair mechanisms, robust antioxidant defenses, and efficient cellular repair machinery safeguard the integrity of reproductive cells, ensuring the fidelity of genetic information passed on to offspring. Additionally, reproductive structures such as spores, cysts, or reproductive buds may exhibit enhanced radiation resistance, allowing for the transmission of viable offspring despite exposure to ionizing radiation.

Resource Allocation Strategies: In radiation-rich environments where resources may be limited or fluctuate unpredictably, Radiotrophic organisms may exhibit adaptive resource allocation strategies to optimize reproductive success. Energy reserves stored during periods of low radiation exposure may be allocated strategically to support reproductive efforts during periods of increased radiation levels, ensuring the continuation of the life cycle despite environmental challenges. Flexible reproductive strategies, including facultative sexual reproduction in response to favorable conditions, may also contribute to the resilience and adaptability of Radiotrophic populations.

Life Cycle Plasticity: Radiotrophic organisms may exhibit life cycle plasticity, allowing for phenotypic flexibility in response to environmental cues, including radiation levels. Phenotypic plasticity in reproductive traits, such as reproductive timing, fecundity, and

reproductive investment, enables Radiotrophic organisms to adjust their reproductive strategies dynamically in response to fluctuating radiation conditions, maximizing reproductive success and population persistence in radiation-rich environments.

Ecological Role

Radiotrophs occupy a niche in their environment as primary producers, similar to plants on Earth. They form the base of the food chain, serving as a vital energy source for higher trophic levels in the ecosystem. Other organisms may have symbiotic relationships with Radiotrophs, relying on them for energy or protection from radiation.

The ecological role of Radiotrophic organisms encompasses their interactions within ecosystems and their influence on ecosystem dynamics, nutrient cycling, and community structure in radiation-rich environments.

Primary Producers: Radiotrophic organisms serve as primary producers in radiation-rich ecosystems, playing a foundational role in energy flow and nutrient cycling. By harnessing radiation energy through specialized metabolic pathways, Radiotrophic organisms convert inorganic compounds into organic biomass, serving as a source of energy and nutrients for higher trophic levels in the ecosystem.

Ecosystem Engineers: Radiotrophic organisms can act as ecosystem engineers, modifying their environment and creating microhabitats that support diverse microbial communities and contribute to ecosystem resilience. Their metabolic activities, including radiation absorption and organic matter decomposition, can alter soil chemistry, nutrient availability, and microbial diversity, shaping the structure and function of radiation-rich ecosystems.

Nutrient Cycling: Radiotrophic organisms participate in nutrient cycling processes, including decomposition, mineralization, and nutrient uptake, facilitating the recycling and redistribution of essential elements within radiation-rich ecosystems. Their metabolic activities contribute to the breakdown of organic matter, release of nutrients from dead biomass, and incorporation of nutrients into microbial biomass, sustaining ecosystem productivity and supporting trophic interactions.

Detoxification and Bioremediation: Radiotrophic organisms possess metabolic capabilities that enable them to detoxify and mitigate the effects of radiation and environmental contaminants. Through mechanisms such as radiation absorption, antioxidant production, and enzymatic detoxification pathways, Radiotrophic organisms can reduce the bioavailability and toxicity of radioactive and chemical pollutants, contributing to the bioremediation of contaminated environments and the restoration of ecosystem health.

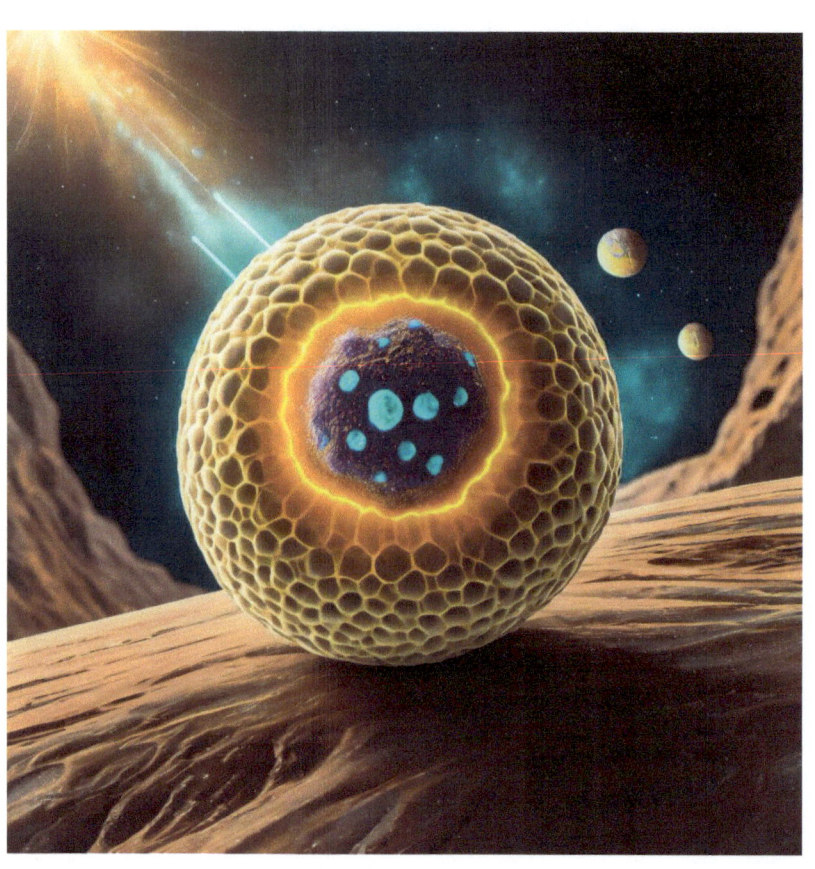

Symbiotic Relationships: Radiotrophs may engage in symbiotic relationships with other organisms, including mutualistic, commensal, or parasitic interactions, depending on the ecological context. Symbiotic associations with other microorganisms, plants, or animals can influence the fitness, distribution, and ecological roles of Radiotrophic organisms, shaping the dynamics of radiation-rich ecosystems and contributing to overall ecosystem functioning.

Ecosystem Resilience: Radiotrophic organisms play a role in maintaining ecosystem resilience and stability in radiation-rich environments by contributing to ecosystem functioning, nutrient cycling, and biodiversity. Their metabolic activities, interactions with other organisms, and adaptive responses to environmental stressors contribute to the overall resilience of radiation-rich ecosystems, enabling them to persist and function despite challenging conditions.

Biogeochemical Cycling: Radiotrophic organisms contribute to biogeochemical cycling processes, including the transformation and cycling of elements such as carbon, nitrogen, and sulfur, within radiation-rich ecosystems. Their metabolic activities influence the availability and cycling of these elements, shaping ecosystem processes and influencing the biogeochemical dynamics of radiation-contaminated environments.

Overall, Radiotrophs play diverse and critical roles in radiation-rich ecosystems, influencing ecosystem structure, functioning, and resilience through their metabolic activities, interactions with other organisms, and contributions to nutrient cycling and biogeochemical processes. Understanding their ecological roles is essential for understanding the dynamics of radiation-contaminated environments.

Evolutionary History
The evolution of Radiotrophs is likely influenced by the intense radiation levels in their environment. Over time, they have undergone adaptive radiation, diversifying into various specialized forms to exploit different radiation sources and niches within their ecosystem.

Transitioning from inorganic to organic matter, would involve a series of complex biochemical and evolutionary steps.

Prebiotic Environment: The hypothetical Radiotrophs originate in a prebiotic environment rich in inorganic compounds, including water (possibly), minerals, gases (such as hydrogen, methane, and ammonia), and cosmic radiation. These conditions provide the necessary raw materials and energy sources for chemical reactions to occur.

Radiation-Driven Chemistry: In this radiation-rich environment, ionizing radiation interacts with inorganic molecules, leading to the formation of simple organic compounds through processes such as radiolysis and photochemical reactions. These organic molecules include amino acids, nucleotides, sugars, and lipid precursors.

Formation of Protocells: As organic molecules accumulate, they self-assemble into protocellular structures, driven by hydrophobic interactions, electrostatic forces, and chemical gradients. These protocells are primitive, membrane-bound compartments capable of encapsulating and concentrating organic molecules, providing a conducive environment for further chemical reactions.

Emergence of Radiotrophic Metabolism: Within these protocells, certain organic molecules with metal cofactors exhibit unique properties that allow them to absorb and utilize ionizing radiation as a source of energy. This early form of Radiotrophic metabolism involves simple biochemical pathways capable of harnessing radiation energy to drive rudimentary cellular processes, such as membrane transport and molecular synthesis.

Natural Selection and Evolution: Protocells with Radiotrophic metabolism gain a selective advantage in the prebiotic environment due to their ability to exploit radiation as an energy source. Through natural selection, variants with more efficient radiation-absorbing structures and metabolic pathways outcompete others, leading to the gradual evolution of more sophisticated Radiotrophic organisms.

Genetic Complexity and Adaptation: Over time, Radiotrophic organisms acquire genetic complexity through mechanisms such as RNA-based replication, error-prone replication, and natural selection. This genetic complexity allows for the refinement and diversification of Radiotrophic metabolic pathways, as well as the development of mechanisms for DNA repair, regulation, and adaptation to changing environmental conditions.

Emergence of Cellular Complexity: As Radiotrophic organisms evolve, they undergo transitions from simple protocellular structures to more complex cellular forms with specialized organelles, membranes, and metabolic compartments. This cellular complexity enables more efficient energy capture, storage, and utilization, potentially leading to the emergence of multicellular Radiotrophic organisms with differentiated cell types and tissues.

Adaptation to Diverse Environments: Radiotrophic organisms adapt to diverse environments characterized by varying levels of radiation, temperature, pH, and nutrient availability. Through genetic diversity and phenotypic plasticity, they colonize different habitats, including deep space, planetary surfaces, subsurface environments, and extreme environments, where radiation serves as a primary energy source.

Radiotrophs may be an entirely new organic entity and may or may not be similar to viruses or even be of the known, earth-based, three-domain system that divides cellular life forms into three domains, namely Archaea, Bacteria, and Eukarya. Additionally, viruses are not considered part of the three-domain system that divides cellular life forms based on phylogenetic analysis of ribosomal RNA sequences and classifies cellular organisms into three distinct domains based on their genetic and evolutionary relationships.

Bacteria (Domain Bacteria): This domain comprises prokaryotic organisms with distinct cellular characteristics, including lack of membrane-bound organelles, circular chromosomes, and unique cell wall compositions. Bacteria encompass a diverse range of organisms, including familiar bacteria such as Escherichia coli and Bacillus subtilis.

Archaea (Domain Archaea): Archaea are also prokaryotic organisms, but they exhibit distinct genetic, biochemical, and physiological characteristics that differentiate them from bacteria. Archaea thrive in extreme environments such as hot springs, salt flats, and deep-sea hydrothermal vents. Examples of archaea include extremophiles like Halobacterium and Methanopyrus.

Eukarya (Domain Eukarya): Eukarya comprises all organisms with eukaryotic cells, characterized by membrane-bound organelles, linear chromosomes, and a complex cellular structure. This domain includes a wide diversity of organisms, including protists, fungi, plants, and animals.

Viruses, on the other hand, are acellular entities that are not classified within the three-domain system because they do not possess the characteristics of cellular life forms. Viruses are composed of genetic material (either DNA or RNA) enclosed in a protein coat (capsid). They lack cellular machinery for metabolism, growth, and reproduction and rely on host cells to replicate. As such, viruses are typically considered non-cellular entities and are not included in the classification of cellular life forms into the three domains of life.

Earth-based Analogs

Radiotrophic fungi, such as those observed in the aftermath of the Chernobyl nuclear accident, are intriguing examples of organisms capable of thriving in high-radiation environments.

Radiation Adaptation: Radiotrophic fungi possess unique adaptations that enable them to withstand and even *utilize* ionizing radiation as an energy source. These adaptations include mechanisms to repair DNA damage caused by radiation, as well as protective mechanisms to minimize radiation-induced cellular damage.

Melanin Pigmentation: One of the key adaptations of Radiotrophic fungi is the production of melanin, a dark pigment that absorbs radiation. Melanin serves as a natural sunscreen, protecting the fungal cells from the harmful effects of ionizing radiation by absorbing and dissipating its energy.

Radiation Energy Utilization: Melanin not only provides protection from radiation but also serves as a potential energy source for Radiotrophic fungi. Studies have shown that certain fungi, including species of the genus Cryptococcus and Cladosporium, can utilize radiation energy absorbed by melanin to drive metabolic processes, similar to how plants use sunlight in photosynthesis.

Metabolic Pathways: Radiotrophic fungi have evolved specialized metabolic pathways to harness the energy derived from radiation absorbed by melanin. These pathways involve the conversion of radiation energy into chemical energy, which is used to drive essential cellular processes such as growth, reproduction, and maintenance.

Radiation Resistance Mechanisms: In addition to melanin pigmentation, Radiotrophic fungi possess other radiation resistance mechanisms that allow them to survive in high-radiation environments. These mechanisms include efficient DNA repair systems, antioxidant defenses against reactive oxygen species generated by radiation, and structural adaptations to withstand radiation-induced cellular damage.

Ecological Role: Earth-based Radiotrophic fungi play important ecological roles in radiation-contaminated environments such as those near nuclear accidents. They participate in the decomposition of organic matter, nutrient cycling, and soil formation, contributing to ecosystem resilience and recovery in these challenging environments.

Biotechnological Applications: The unique adaptations of Radiotrophic fungi have potential biotechnological applications, particularly in the field of bioremediation of radioactive waste and contaminated environments. These fungi can be used to degrade organic contaminants, immobilize radionuclides, and facilitate the cleanup of radioactive sites.

Radiotrophic fungi exhibit remarkable adaptations to high-radiation environments, including melanin pigmentation, specialized metabolic pathways, and radiation resistance mechanisms. Understanding the biology of these real-world, earth-based organisms not only sheds light on their ecological roles but also holds promise for the potential deep-space cases presented herein.

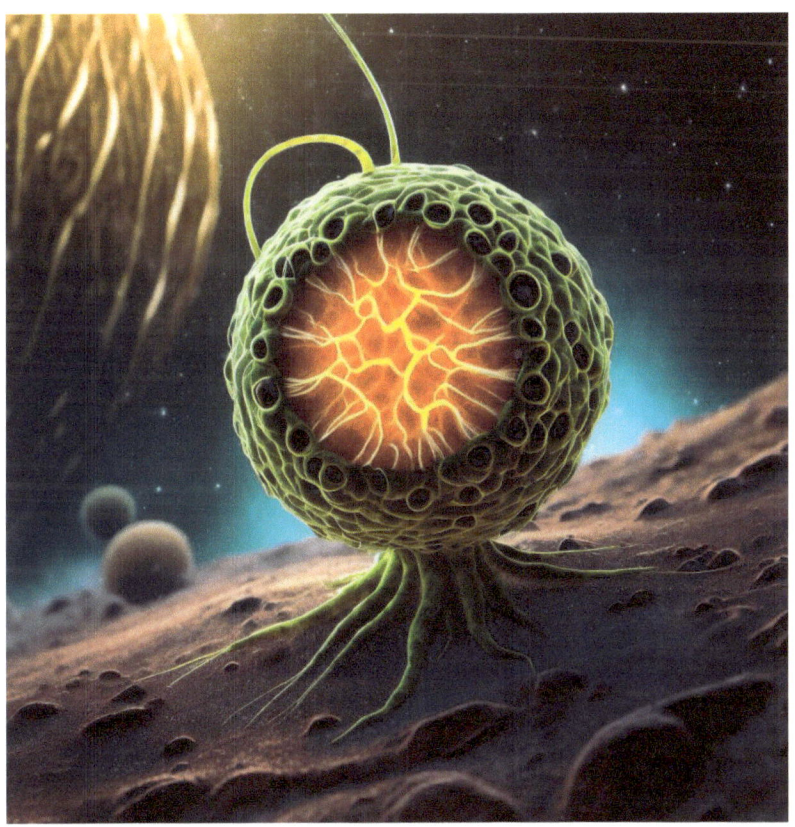

Besides certain fungi like those found in the Chernobyl Exclusion Zone, there are other earth-based organisms known to survive and even thrive in high-radiation environments.

Deinococcus radiodurans: Often referred to as the "conan of bacteria" due to its remarkable radiation resistance, Deinococcus radiodurans is a species of extremophilic bacteria known for its ability to withstand high levels of ionizing radiation, desiccation, and oxidative stress. It is found in diverse terrestrial environments, including soil, water, and food, and has been studied extensively for its applications in bioremediation and astrobiology.

Cryptococcus neoformans: Certain strains of the yeast Cryptococcus neoformans exhibit radiation resistance and can survive in environments with elevated radiation levels. Studies have shown that some strains of Cryptococcus neoformans can utilize melanin pigmentation to protect against radiation-induced damage and may potentially play a role in radiation-contaminated environments.

Thermococcus gammatolerans: Thermococcus gammatolerans is a species of archaea prokaryote found in deep-sea hydrothermal vents characterized by high levels of radiation due to natural radioactivity in the Earth's crust. This extremophilic archaeon exhibits exceptional radiation resistance and has adapted to thrive in the extreme conditions of deep-sea hydrothermal environments, including high temperatures, pressure, and radiation levels.

Halobacterium salinarum: Halobacterium salinarum is a halophilic archaeon found in hypersaline environments such as salt flats, salt lakes, and salt mines. Some strains of Halobacterium salinarum have been shown to exhibit radiation resistance and are capable of surviving exposure to high levels of ionizing radiation. These extremophilic archaea have adapted to thrive in environments with elevated radiation levels and extreme salinity.

Ramellogammarus similis: Ramellogammarus similis is a species of amphipod crustacean found in the Chernobyl Exclusion Zone and other radiation-contaminated environments. Studies have documented the presence of Ramellogammarus similis in areas with high levels of ionizing radiation, suggesting that this species has adapted to survive and reproduce in radiation-rich habitats.

These organisms demonstrate various adaptations and mechanisms that enable them to withstand, *and thrive,* in high-radiation environments, providing valuable insights into the biology of extremophiles and the potential for life to exist in challenging deep space conditions.

Earth hosts a diverse array of extreme environments where life thrives in conditions that challenge conventional notions of habitability. These environments provide valuable analogs for understanding the potential for deep space life in extreme conditions, including those with high levels of ionizing radiation. Exploring these extreme environments not only sheds light on the adaptations of organisms to extreme conditions but also provides insights into the potential habitats and adaptations of Radiotrophic organisms.

Radioactive Waste Sites: Radioactive waste sites, including nuclear power plants, nuclear test sites, and areas affected by nuclear accidents such as Chernobyl and Fukushima, represent unique environments with elevated levels of ionizing radiation. Despite the harsh conditions, these sites are inhabited by organisms known as radioresistant extremophiles. These extremophiles have evolved mechanisms to repair radiation-induced DNA damage, scavenge free radicals, and survive in the presence of radioactive isotopes. Studying the adaptations of organisms in radioactive waste sites provides valuable insights into the potential adaptations of Radiotrophic organisms to high-radiation environments.

Deep-Sea Hydrothermal Vents: Deep-sea hydrothermal vents are dynamic ecosystems located along mid-ocean ridges where hot, mineral-rich fluids emanate from the seafloor. These environments are characterized by extreme conditions, including high pressure, temperature gradients, and chemically rich fluids. Despite these challenges, deep-sea hydrothermal vents host a variety of extremophiles adapted to thrive in these harsh conditions. These extremophiles, including thermophiles, chemolithotrophs, and piezophiles, utilize unique metabolic pathways to harness chemical energy from the vent fluids. Understanding the adaptations of organisms in deep-sea hydrothermal vents provides insights into the metabolic diversity and survival strategies of extremophiles in extreme environments, including their relevance to Radiotrophic biology.

Subglacial Lakes: Subglacial lakes, such as Lake Vostok in Antarctica, are bodies of liquid water located beneath the Antarctic ice sheet. These environments are characterized by extreme cold, high pressure, and isolation from sunlight and external nutrients. Despite these challenging conditions, subglacial lakes harbor microbial communities adapted to thrive in the dark, nutrient-poor environment. These extremophiles, including psychrophiles and oligotrophs, utilize metabolic strategies to survive and metabolize organic and inorganic compounds under extreme conditions. Studying the adaptations of organisms in subglacial lakes provides insights into the metabolic versatility and survival mechanisms of extremophiles in cold, dark, and isolated environments, with implications for understanding the potential habitats of Radiotrophic organisms in extraterrestrial environments.

Polar Regions: Polar regions, including the Arctic and Antarctic, are characterized by extreme cold, low temperatures, and seasonal fluctuations in light availability. Despite the harsh conditions, polar regions support diverse microbial communities adapted to thrive in cold, nutrient-poor environments. These extremophiles, including psychrophiles, cryophiles, and halophiles, exhibit adaptations to survive and metabolize organic and inorganic compounds under extreme cold and light conditions.

Exploring extreme environments on Earth provides valuable insights into the adaptations of organisms to extreme conditions and their relevance to understanding Radiotrophic biology. Studying the adaptations of organisms in radioactive waste sites, deep-sea hydrothermal vents, subglacial lakes, and polar regions sheds light on the metabolic diversity, survival strategies, and potential habitats of extremophiles, including Radiotrophic organisms, in extreme environments on Earth, and beyond. By studying the adaptations of organisms to extreme conditions, researchers gain insights into the potential for life in diverse environments, including high-radiation environments, and advance our understanding of the limits of life in the universe.

Biotechnological Applications

The unique metabolic capabilities of deep space Radiotrophic organisms hold significant promise for a wide range of biotechnological applications, ranging from environmental remediation to materials science and astrobiology. Exploring these potential applications not only highlights the practical implications of

Radiotrophic biology but also underscores the interdisciplinary nature of research in this field.

Bioremediation of Radioactive Waste: One of the most promising applications of Radiotrophic organisms is in the bioremediation of radioactive waste sites. Radiotrophic organisms possess the ability to metabolize radioactive isotopes and convert them into less harmful forms through various biochemical pathways. By harnessing the metabolic activities of these organisms, researchers may aim to develop bioremediation strategies for cleaning up contaminated environments, including nuclear power plants, nuclear test sites, and areas affected by nuclear accidents. Radiotrophic organisms could be engineered or applied in situ to facilitate the remediation of radioactive waste by reducing the concentration and mobility of radioactive contaminants in soil, water, and sediments.

Development of Radiation-Resistant Materials: Radiotrophic organisms offer valuable insights into the mechanisms of radiation resistance and DNA repair, which can inform the development of radiation-resistant materials for various applications. By studying the molecular mechanisms underlying radiation resistance in Radiotrophic organisms, researchers can identify biomolecules, enzymes, and pathways that confer resistance to ionizing radiation. These insights can be applied to engineer novel materials, such as radiation-resistant polymers, coatings, or biomimetic materials, for use in space exploration, nuclear industries, medical imaging, and other radiation-intensive applications. Radiation-resistant materials inspired by Radiotrophic biology could enhance the durability, safety, and performance of technologies operating in high-radiation environments.

Applications in Astrobiology and Space Exploration: Radiotrophic organisms hold relevance for astrobiology and space exploration due to their potential to survive and metabolize in high-radiation environments analogous to those found in space. By studying the adaptations of Radiotrophic organisms to extreme conditions on Earth, researchers gain insights into the potential habitability of extraterrestrial environments, including regions of cosmic radiation in interstellar space, planetary surfaces, or subsurface habitats. Radiotrophic organisms could serve as model systems for studying the limits of life in space and informing the design of life-detection instruments for future space missions. Furthermore, understanding the metabolic capabilities of Radiotrophic organisms could inform the development of bioregenerative life support systems for long-duration space missions, where radiation exposure is a significant concern.

The biotechnological applications of Radiotrophic organisms span diverse fields, including environmental remediation, materials science, and astrobiology. By harnessing the unique metabolic capabilities of Radiotrophic organisms, researchers aim to develop innovative solutions for cleaning up radioactive waste, engineering radiation-resistant materials, and exploring the potential for life beyond Earth. The study of Radiotrophic biology not only advances our understanding of extremophilic life forms on Earth but also inspires novel approaches to addressing environmental challenges, developing advanced materials, and exploring the potential for life in extreme environments in the universe.

Non-Water-based Life Chemistry

Water is generally considered a crucial factor for life as we know it, including the majority of terrestrial life forms. Given the importance of water to terrestrial life, this section is dedicated to alternative chemistry scenarios for deep space alien Radiotrophic life.

Water serves as a universal solvent, facilitating biochemical reactions and supporting cellular processes such as metabolism, transport, and structure. It also provides a medium for the dissolution and transport of nutrients and essential molecules within organisms. Additionally, water plays a vital role in stabilizing biological molecules and maintaining cellular homeostasis through processes like osmoregulation.

In the context of astrobiology and the search for life beyond Earth, the presence of liquid water is often considered a key factor in determining the potential habitability of a planetary body. This is because water is abundant in the universe, exists in liquid form over a wide range of temperatures, and supports a diverse array of chemical reactions necessary for life.

It is reasonable to speculate that water could be essential for the functioning of Radiotrophic organisms, particularly if they share biochemical similarities with known life forms. However, it's also conceivable that in extreme environments where liquid water is scarce or absent, *alternative solvents or adaptations* could support Radiotrophic life. While it is challenging to find the following proposed solvents in liquid form, versus their more common gaseous state, it is entirely feasible given the vastness of the universe and the immense number of exoplanets and cometary bodies.

Ammonia

One alternative solvent that has been proposed as a potential medium for life in extreme environments is liquid ammonia (NH_3). Liquid ammonia has several properties that make it an intriguing candidate for supporting life in environments where water is scarce or absent.

Solvent Properties: Like water, liquid ammonia is a polar*(footnote) solvent capable of dissolving a wide range of organic and inorganic compounds. It can facilitate biochemical reactions and support the transport of essential molecules within organisms.

Stability: Liquid ammonia remains in a liquid state over a wide temperature range, with its melting point significantly lower than that of water (-77.7°C) and its boiling point higher (-33.3°C at atmospheric pressure). This stability could allow for the existence of liquid ammonia environments on deep space planetary bodies with extreme temperatures.

Hydrogen Bonding: Ammonia molecules can form hydrogen bonds with other molecules, although weaker than those formed by water molecules. This property could contribute to the stabilization of biological molecules and the maintenance of cellular structures in ammonia-based Radiotrophic life forms.

Chemical Reactivity: Ammonia is chemically reactive and can participate in various chemical reactions relevant to metabolism and energy production. It can serve as a source of nitrogen (N_2), an essential element for life, and participate in redox reactions similar to those involving water in traditional biochemistry.

Potential Habitats: WISE 1828+2650 is a binary brown dwarf, or rogue planet. A team of researchers (members from the Institute for Particle Physics and Astrophysics at ETH Zurich) using the James Webb Space Telescope (JWST), with its Mid-Infrared Instrument (MIRI), detected water vapor (H_2O), methane (CH_4) and two isotopologues of ammonia in the atmosphere of WISE 1828+2650. WISE 1828+2650 is 32.5 light years from Earth, located in the constellation Lyra, the lyre. Brown dwarfs are somewhere in between stars and planets: they resemble giant gas planets in many ways, which is why they can be used as a model system to study gas giants.

Another notable example is the exoplanet HD 189733b, which is a hot-Jupiter located about 64 light-years away from Earth. Spectroscopic observations of

HD 189733b's atmosphere have revealed the presence of various molecules, including water vapor, carbon dioxide, and possibly ammonia. The detection of ammonia in this exoplanet's atmosphere has been inferred from the analysis of absorption features in its spectrum.

Another example is the exoplanet WASP-33b, which is also a hot-Jupiter located about 380 light-years away from Earth. Observations of WASP-33b's atmosphere have suggested the presence of ammonia based on the analysis of its thermal emission spectrum.

Ammonia has been detected in various regions of deep space, particularly in interstellar clouds and molecular clouds. These clouds are vast regions of gas and dust where new stars and planetary systems are forming. Ammonia is one of the many molecules that have been identified through spectroscopic observations of these clouds.

Interstellar clouds are composed of gas and dust, primarily hydrogen and helium, along with traces of other elements and molecules. In these clouds, ammonia can form through chemical reactions involving nitrogen and hydrogen molecules, as well as other atomic and molecular species. The detection of ammonia in interstellar clouds provides insights into the chemical composition and processes occurring in these regions.

Molecular clouds are denser regions within interstellar clouds where star formation occurs. These clouds contain higher concentrations of molecules, including ammonia, which serves as a precursor to more complex organic molecules. Observations of molecular clouds have revealed the presence of ammonia through its characteristic spectral lines in the radio and infrared regions of the electromagnetic spectrum.

Overall, the detection of ammonia in interstellar and molecular clouds demonstrates the widespread distribution of this molecule in deep space and its role in the chemistry of star-forming regions.

Where liquid water is scarce or absent, liquid ammonia has been proposed as a potential solvent for supporting alternative forms of life, sometimes referred to as "ammonia-based life" or "ammonia-based biochemistry." While such hypothetical life forms, including Radiotrophic organisms, have not been observed or studied, the properties of liquid ammonia suggest it could provide a viable medium for Radiotrophic life under extreme environmental conditions.

Methane
Another solvent that has been proposed as a potential candidate to support hypothetical Radiotrophic life on a distant planet is liquid methane (CH_4). While methane is typically found in a gaseous state on Earth, it can exist as a liquid under certain conditions of low temperature and high pressure, such as those found on the surface of distant, cold planets or moons in our solar system and beyond.

Solvent Properties: Liquid methane is non-polar and can dissolve non-polar organic molecules. While it has different solvent properties compared to water or ammonia, it could still facilitate biochemical reactions and support the transport of essential molecules within organisms.

Stability: Liquid methane remains in a liquid state under conditions of low temperature and high pressure, similar to those found on distant, cold planetary bodies. Its stability as a liquid solvent under these conditions could allow for the existence of methane-based environments.

Chemical Reactivity: Methane molecules can participate in various chemical reactions, including redox reactions and hydrocarbon chemistry. While its chemical reactivity differs from that of water or ammonia, methane could still serve as a source of carbon and energy for potential methane-based Radiotrophic life forms.

Potential Habitats: Planetary bodies such as Enceladus & Titan, two moons of Saturn, have been suggested as potential habitats for methane-based life due to the presence of liquid methane lakes and rivers on their surfaces. While their conditions are extremely cold, the presence of liquid methane and other hydrocarbons raises the possibility of methane-based life forms adapted to these conditions. Methane has also been detected in other astronomical environments, such as in the atmospheres of planets and moons in our own solar system, including Jupiter, Uranus and Neptune. The presence of methane in these environments provides insights into their atmospheric compositions and processes.

Methane has been detected in various astronomical environments, including exoplanets and regions of deep space through spectroscopic observations. One notable example is the exoplanet HD 189733b, a hot-Jupiter located about 64 light-years away from Earth in the constellation of Vulpecula. Methane has been inferred in the atmosphere of HD 189733b, along with other molecules like water vapor and carbon dioxide.

NASA's James Webb Space Telescope observed the warm-Jupiter exoplanet WASP-80b (aka Wadirum) as it passed in front of and behind its host star, revealing spectra indicative of an atmosphere containing methane gas and water vapor. WASP-80 is a K-type main-sequence star about 162 light-years away. While water vapor has been detected in over a dozen planets to date, until recently, methane, a molecule found in abundance in the atmospheres of Jupiter, Saturn, Uranus, and Neptune within our solar system, has remained elusive in the atmospheres of transiting exoplanets when studied with space-based spectroscopy.

Methane has also been detected in interstellar clouds and molecular clouds, which are vast regions of gas and dust where new stars and planetary systems are forming. These clouds contain a diverse range of molecules, including methane, which is formed through chemical reactions involving hydrogen and carbon-bearing species like carbon monoxide and carbon dioxide. Observations of these clouds have revealed the presence of methane through its characteristic spectral lines in the radio and infrared regions of the electromagnetic spectrum.

Overall, the detection of methane in exoplanets and regions of deep space highlights its widespread distribution in astronomical environments and its role in astrochemistry and planetary science. Studying the abundance and distribution of methane and other molecules in these environments contributes to our understanding of the chemical evolution of the universe and the formation and evolution of planetary systems.

Ethane
Another potential candidate to support hypothetical Radiotrophic life on a distant planet is liquid ethane (C_2H_6). Like methane, ethane is typically found in a gaseous state on Earth but can exist as a liquid under certain conditions of low temperature and high pressure, such as those found on the surface of distant, cold planets or moons in our solar system and beyond.

Solvent Properties: Liquid ethane is non-polar and can dissolve non-polar organic molecules. While it has different solvent properties compared to water, ammonia, or methane, it could still facilitate biochemical reactions and support the transport of essential molecules within organisms.

Stability: Liquid ethane remains in a liquid state under conditions of low temperature and high pressure, similar to those found on distant, cold planetary bodies. Its stability as a liquid solvent under these conditions could allow for the existence of ethane-based environments.

Chemical Reactivity: Ethane molecules can participate in various chemical reactions, including redox reactions and hydrocarbon chemistry. While its chemical reactivity differs from that of water, ammonia, or methane, ethane could still serve as a source of carbon and energy for potential ethane-based Radiotrophic life forms.

Potential Habitats: Planetary bodies with cold, hydrocarbon-rich environments, such as Titan and other moons of gas giants, have been suggested as potential habitats for ethane-based life due to the presence of liquid ethane and other hydrocarbons on their surfaces.

Ethane has been detected in various astronomical environments, including exoplanets and regions of deep space, through spectroscopic observations, including the previously mentioned exoplanet HD 189733b, as well as cometary atmospheres, interstellar clouds and molecular clouds.

One notable example is Comet 67P/Churyumov-Gerasimenko. This comet, also known as "Rosetta's Comet," was the target of the European Space

Agency's Rosetta mission, which orbited the comet and deployed the Philae lander to its surface in 2014.

The presence of ethane in these environments provides insights into atmospheric compositions and processes and highlights its widespread distribution in astronomical environments.

Hydrogen

Another potential solvent is liquid hydrogen. Liquid hydrogen is a molecular hydrogen in a liquid state, which requires extremely low temperatures (below its boiling point of -252.87°C) to maintain.

Solvent Properties: Liquid hydrogen is non-polar and can dissolve non-polar organic molecules. While it has different solvent properties compared to water, ammonia, methane, or ethane, it could still potentially facilitate biochemical reactions and support the transport of essential molecules within organisms.

Stability: Liquid hydrogen remains in a liquid state under extremely low temperatures, similar to those found in the outer reaches of the solar system or on cold, distant planetary bodies. Its stability as a liquid solvent under these conditions could allow for the existence of hydrogen-based environments.

Chemical Reactivity: Hydrogen molecules can participate in various chemical reactions, including redox reactions and hydrogenation reactions. While its chemical reactivity differs from that of other solvents, liquid hydrogen could still potentially serve as a source of energy for hypothetical hydrogen-based Radiotrophic life forms.

Potential Habitats: Planetary bodies with extremely cold environments, such as the outer moons of gas giants or certain regions in the outer reaches of the solar system, could potentially harbor liquid hydrogen environments. While the conditions in these environments are harsh and challenging, the presence of liquid hydrogen raises the possibility of hydrogen-based life forms adapted to these conditions.

Hydrogen is the most abundant element in the universe and is found in a wide range of astronomical environments, including exoplanets and regions of deep space.

Hydrogen is a major component of the atmospheres of many exoplanets and gas giants like Jupiter and Saturn, as well as hot-Jupiters and hot-Neptunes. Spectroscopic observations of exoplanet atmospheres have revealed the presence of hydrogen through its characteristic absorption and emission features in the ultraviolet, visible, and infrared regions of the electromagnetic spectrum. Hydrogen is also inferred to be present in the atmospheres of rocky exoplanets, although it may be in the form of water vapor or hydrocarbons rather than molecular hydrogen.

It is abundant in interstellar clouds and molecular clouds, which are vast regions of gas and dust where new stars and planetary systems are forming. These clouds are primarily composed of molecular hydrogen (H_2), along with atomic hydrogen (H) and other molecules. Observations of these clouds have revealed the presence of hydrogen through its characteristic spectral lines in the radio, microwave, and infrared regions of the electromagnetic spectrum. Molecular hydrogen is particularly abundant in cold, dense regions of interstellar clouds, where it serves as the primary fuel for star formation.

WASP-121b: This exoplanet is a hot-Jupiter located approximately 880 light-years from Earth in the constellation of Puppis. WASP-121b is known for its extreme temperature and proximity to its host star. Observations suggest that its atmosphere contains significant amounts of gaseous hydrogen, as well as other molecules like water vapor and helium.

HD 209458b: This exoplanet, also known as Osiris, is a hot-Jupiter located approximately 157 light-years from Earth in the constellation of Pegasus. HD 209458b is one of the most studied exoplanets and has been found to have a hydrogen-dominated atmosphere, along with traces of other elements and molecules such as water vapor, carbon monoxide, and carbon dioxide.

HD 189733b: This exoplanet is another hot-Jupiter located approximately 64 light-years away in the constellation of Vulpecula. HD 189733b has been extensively studied, and while hydrogen is not directly detected in its atmosphere, it is inferred to be a major constituent based on models and observations of other hot-Jupiters with similar characteristics.

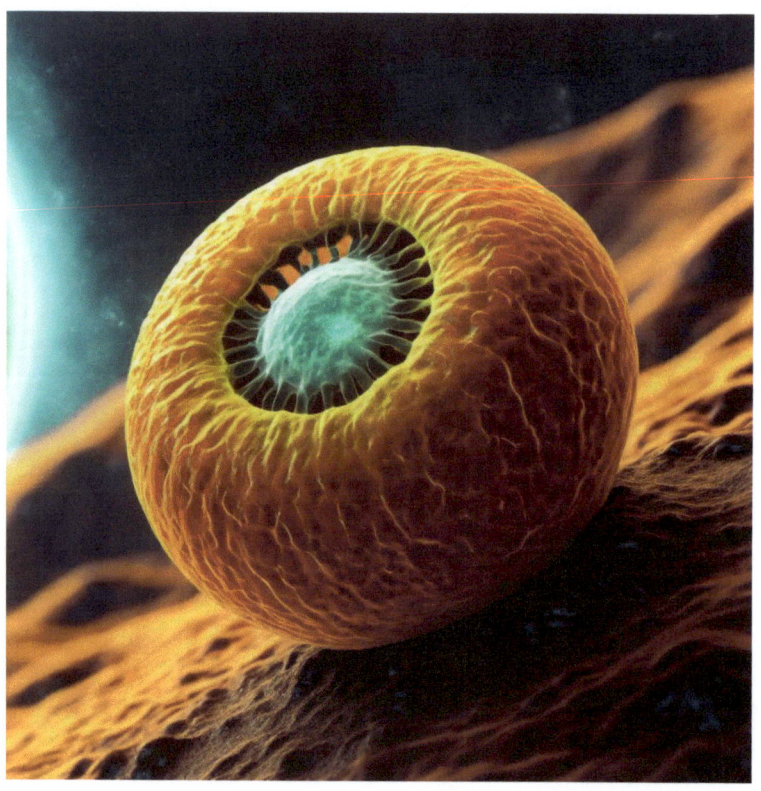

Sulfur Dioxide

Liquid sulfur dioxide (SO_2) is a molecular compound that can exist in a liquid state under specific temperature and pressure conditions.

Solvent Properties: Liquid sulfur dioxide is a polar solvent capable of dissolving polar and non-polar organic molecules. While it has different solvent properties compared to water, ammonia, methane, ethane, or hydrogen, it could potentially facilitate biochemical reactions and support the transport of essential molecules within organisms.

Stability: Liquid sulfur dioxide remains in a liquid state under conditions of moderate temperature and pressure, which could be present on certain planetary bodies with unique atmospheric compositions or geological processes. Its stability as a liquid solvent under these conditions could allow for the existence of sulfur dioxide-based environments.

Chemical Reactivity: Sulfur dioxide molecules can participate in various chemical reactions, including redox reactions and acid-base reactions. While its chemical reactivity differs from that of other solvents, liquid sulfur dioxide could potentially serve as a source of energy for hypothetical sulfur dioxide-based Radiotrophic life forms.

Potential Habitats: Sulfur dioxide has been detected in various interstellar and planetary nebulae, which are regions of ionized gas and dust surrounding stars. These nebulae contain a variety of molecules, including sulfur dioxide, which is formed through chemical processes involving sulfur-bearing species and oxygen. Observations of these nebulae have revealed the presence of sulfur dioxide through its characteristic spectral lines in the radio and infrared regions of the electromagnetic spectrum.

While not a direct observation of sulfur dioxide in deep space, Io, one of Saturn's moons, is known for its volcanic activity and sulfur-rich environment. Observations by spacecraft such as NASA's Galileo mission have revealed plumes of sulfur dioxide and other sulfur-bearing compounds erupting from volcanoes on Io's surface.

Astronomers have used 2023 observations made with the JWST to study the atmosphere of the nearby exoplanet WASP-107b where they discovered not only water vapor and sulfur dioxide, but even silicate sand clouds. WASP-107b is a super-Neptune ice giant exoplanet that orbits the star WASP-107, 200 light-years away from Earth in the constellation Virgo.

In 2022, JWST's array of highly sensitive instruments was trained on the atmosphere of WASP-39b (aka Bocaprins), a hot-Saturn orbiting a star some 700 light-years away and identified sulfur dioxide in the exoplanet's atmosphere. Its presence can only be explained by photochemistry, chemical reactions triggered by high-energy particles of starlight.

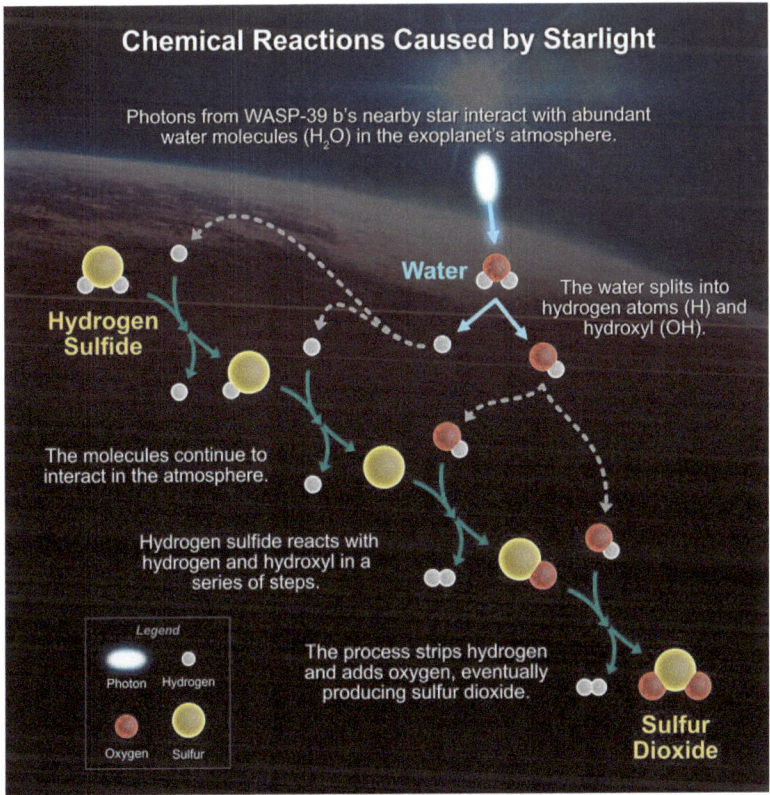

Chemical Reactions Caused by Starlight

Photons from WASP-39 b's nearby star interact with abundant water molecules (H_2O) in the exoplanet's atmosphere.

Water

The water splits into hydrogen atoms (H) and hydroxyl (OH).

Hydrogen Sulfide

The molecules continue to interact in the atmosphere.

Hydrogen sulfide reacts with hydrogen and hydroxyl in a series of steps.

Legend

Photon Hydrogen

Oxygen Sulfur

The process strips hydrogen and adds oxygen, eventually producing sulfur dioxide.

Sulfur Dioxide

Image Credit: NASA/JPL - Caltech/Robert Hurt; and the Center for Astrophysics - Harvard & Smithsonian, Melissa Weiss

Nitrogen

Nitrogen (N_2) can exist as a liquid under certain conditions of extremely low temperature and high pressure, such as those found in the outer reaches of the solar system or on cold, distant planetary bodies.

Solvent Properties: Liquid nitrogen is non-polar and can dissolve non-polar organic molecules. While it has different solvent properties compared to water, ammonia, methane, ethane, hydrogen, or sulfur dioxide, it could potentially facilitate biochemical reactions and support the transport of essential molecules within organisms.

Stability: Liquid nitrogen remains in a liquid state under extremely low temperatures, similar to those found in the outer reaches of the solar system or on cold, distant planetary bodies. Its stability as a liquid solvent under these conditions could allow for the existence of nitrogen-based environments.

Chemical Reactivity: Nitrogen molecules can participate in various chemical reactions, including redox reactions and nitrogen fixation. While its chemical reactivity differs from that of other solvents, liquid nitrogen could potentially serve as a source of energy for hypothetical nitrogen-based Radiotrophic life forms.

Potential Habitats: Planetary bodies with extremely cold environments, such as certain regions in the outer reaches of the solar system or distant exoplanets with nitrogen-rich atmospheres, could potentially harbor liquid nitrogen environments. While the conditions in these environments are harsh and challenging, the presence of liquid nitrogen raises the possibility of nitrogen-based life forms adapted to these conditions.

Nitrogen is a fundamental component of many astronomical environments, but specific detections of molecular nitrogen in its diatomic form are less common compared to other molecules like ammonia or nitrogen-containing organic compounds.

Titan, the largest moon of Saturn, is known for its dense atmosphere composed primarily of nitrogen with smaller amounts of methane and other hydrocarbons. The Cassini-Huygens mission provided detailed observations of Titan's atmosphere, revealing its nitrogen-rich composition.

Measurements show that planetary nebula NGC 6153, 4000 light-years from Earth in the constellation Scorpius, contains large amounts of neon, argon, oxygen, carbon and chlorine, up to three times more than can be found in the solar system. This nebula also contains five times more nitrogen than our sun. Although it may be that it developed higher levels of these elements as it grew and evolved, it is more likely that it originally formed from a cloud of material that already contained significantly more of these elements.

The Japan Aerospace Exploration Agency's (JAXA) Hayabusa2 spacecraft has explored the near-Earth C-type carbonaceous asteroid Ryugu and brought back surface material containing iron nitride on magnetite grains from the surface of Ryugu.

Isotope analyses of lunar ilmenite grains have revealed that the supply of exogenous nitrogen to the lunar surface is from asteroidal micrometeoroids infalling to the Moon. Exogenous nitrogen could become trapped in lunar regolith by chemisorption or by the formation of nitride with silicon or titanium.

While direct detections of molecular nitrogen in exoplanet atmospheres are challenging, models and observations have suggested the potential presence of nitrogen-bearing compounds in certain exoplanet atmospheres.

Carbon Dioxide
Carbon dioxide (CO_2) can exist as a liquid under specific conditions of low temperature and high pressure, such as those found in certain regions of outer space or on cold, distant planetary bodies.

Solvent Properties: Liquid carbon dioxide is non-polar and can dissolve non-polar organic molecules. While it has different solvent properties compared to water, ammonia, methane, ethane, hydrogen, sulfur dioxide, or nitrogen, it could potentially facilitate biochemical reactions and support the transport of essential molecules within organisms.

Stability: Liquid carbon dioxide remains in a liquid state under conditions of low temperature and high pressure, similar to those found in certain regions of outer space or on cold, distant planetary bodies. Its stability as a liquid solvent under these conditions could allow for the existence of carbon dioxide-based environments.

Chemical Reactivity: Carbon dioxide molecules can participate in various chemical reactions, including carbonation reactions and acid-base reactions. While its chemical reactivity differs from that of other solvents, liquid carbon dioxide could potentially serve as a source of carbon and energy for hypothetical carbon dioxide-based Radiotrophic life forms.

Potential Habitats: Planetary bodies with cold, carbon dioxide-rich atmospheres or surface conditions, such as certain regions of outer space or distant exoplanets with carbon dioxide-rich environments, could potentially harbor liquid carbon dioxide environments. While the conditions in these environments may be extreme, the presence of liquid carbon dioxide raises the possibility of carbon dioxide-based life forms adapted to these conditions.

A new investigation with NASA's JWST into K2-18b, an exoplanet 2.6 times Earth's radius, has revealed the presence of carbon-bearing molecules including methane and carbon dioxide. Webb's discovery adds to recent studies suggesting that K2-18b could be a Hycean exoplanet, one which has the potential to possess a hydrogen-rich atmosphere and a water ocean-covered surface. K2-18b orbits the cool dwarf star K2-18 in the habitable zone and lies 124 light-years from Earth in the constellation Leo.

The abundance of methane and carbon dioxide, and shortage of ammonia, supports the hypothesis that there may be a water ocean underneath a hydrogen-rich atmosphere in K2-18b. These initial Webb observations also provided a possible detection of a molecule called dimethyl sulfide (DMS). *On Earth, this is predominantly only produced by life.* The bulk of the DMS in Earth's atmosphere is emitted from phytoplankton in marine environments.

Others

While water, ammonia, methane, ethane, hydrogen, sulfur dioxide, nitrogen, and carbon dioxide are some of the most commonly discussed solvents in astrobiology for supporting hypothetical life forms, there are other potential solvents that have been proposed in theoretical studies and discussions.

Liquid Hydrocarbons: Solvents such as liquid propane (C_3H_8), liquid butane (C_4H_{10}), or other hydrocarbons could potentially serve as solvents for life forms in environments with specific temperature and pressure conditions, similar to those found on certain moons or exoplanets.

Liquid Fluorine Compounds: Compounds such as liquid fluoromethane (CH_3F) or liquid fluorocarbons have been suggested as potential solvents for life forms in environments with extreme cold temperatures and unique chemical compositions.

Liquid Silicon Compounds: Compounds such as liquid silicones or liquid siloxanes could potentially serve as solvents for life forms in environments with extreme temperatures and pressures, such as those found on certain rocky exoplanets or moons. Researchers found silicate sand clouds residing within the previously mentioned WASP-107b exoplanet's dynamic atmosphere.

Liquid Ammonium Compounds: Compounds such as liquid ammonium hydrosulfide (NH_4SH) or liquid ammonium fluoride (NH_4F) have been proposed as potential solvents for life forms in environments with specific temperature and chemical compositions.

The spectrum of Jupiter's Great Red Spot showed ultraviolet–visible spectra of irradiated ammonium hydrosulfide, a reported Jovian atmospheric cloud component.

Liquid Metal Alloys: Certain liquid metal alloys, such as gallium-indium alloys or other eutectic mixtures, have been suggested as potential solvents for extremophilic life forms in environments with extreme temperatures or chemical conditions.

These are just a few examples of alternative (other than water) solvents that have been proposed in theoretical studies and discussions within the field of astrobiology. Each of these solvents presents unique challenges and opportunities for supporting hypothetical life forms, and further research is needed to explore their potential suitability and the possible adaptations of Radiotrophic life forms to these environments.

* In chemistry, the terms "polar" and "nonpolar" refer to the distribution of electrical charge within a molecule, which affects its interactions with other molecules and solvents. These terms are used to describe the overall polarity of a molecule based on its molecular structure and the arrangement of its atoms.

Polar Molecules: Polar molecules have an uneven distribution of electrical charge, resulting in a separation of positive and negative charges within the molecule. This occurs when there is an asymmetrical distribution of atoms and/or electronegativity differences between atoms within the molecule. Polar molecules have a permanent dipole moment, meaning they have a positive end and a negative end. Examples of polar molecules include water (H_2O), ammonia (NH_3), and hydrogen fluoride (HF).

Nonpolar Molecules: Nonpolar molecules have an even distribution of electrical charge, with no separation of positive and negative charges within the molecule. This occurs when there is a symmetrical distribution of atoms and/or similar electronegativity between atoms within the molecule. Nonpolar molecules do not have a permanent dipole moment. Examples of nonpolar molecules include methane (CH_4), carbon dioxide (CO_2), and diatomic gases like nitrogen (N_2) and oxygen (O_2).

The polarity of a molecule influences its interactions with other molecules and solvents. Polar molecules tend to interact strongly with other polar molecules or ions through electrostatic attractions (such as hydrogen bonding), while nonpolar molecules tend to interact with other nonpolar molecules through weaker dispersion forces (such as van der Waals forces). Understanding the polarity of molecules is important in various fields of chemistry, including organic chemistry, biochemistry, and materials science.

Further Research
There are various earth-based models and analogs that researchers can use to study aspects related to Radiotrophic organisms and their adaptations. While these models may not directly replicate the extreme radiation conditions of deep space or nuclear disaster sites, they provide valuable insights into the mechanisms and ecological implications of radiation biology. Some of these models include:

Radiation-Contaminated Environments: Researchers study natural and anthropogenically contaminated environments, such as uranium mines, nuclear accident sites (e.g., Chernobyl and Fukushima), and areas with high levels of natural background radiation (e.g., Ramsar in Iran). These environments serve as natural laboratories to investigate the effects of radiation on organisms, adaptation strategies, and ecosystem responses.

Radiation-Resistant Microorganisms: Certain extremophilic microorganisms found in terrestrial environments exhibit high levels of radiation resistance and metabolic diversity. Examples include Deinococcus radiodurans, which can withstand extreme levels of ionizing radiation and oxidative stress, and extremophilic fungi and bacteria inhabiting radioactive waste repositories and radionuclide-contaminated sites. In fact, in Ramsar Iran, (the most radioactive inhabited area known on Earth), early anecdotal evidence from local doctors and preliminary cytogenetic studies suggested that there may be no increased incidence of human radiation-induced cancer, and possibly even a *radio-adaptive effect.*

Mars Analog Environments: Mars analog environments on Earth, such as deserts, polar regions, and volcanic terrains, are used to simulate conditions similar to those found on the Martian surface. These environments experience elevated levels of radiation due to reduced atmospheric shielding and can provide insights into the potential survival strategies of radiation-resistant organisms in extraterrestrial habitats.

Astrobiology Research Facilities: Specialized astrobiology research facilities and laboratories simulate extraterrestrial conditions, including radiation-rich environments, to study the survivability and adaptability of organisms under extreme conditions. These facilities utilize radiation sources, vacuum chambers, and controlled environments to mimic space conditions and investigate the potential for life beyond Earth.

Laboratory Experiments: Laboratory experiments using controlled radiation exposure conditions allow researchers to study the effects of ionizing radiation on various organisms, including model organisms such as yeast, nematodes, fruit flies, and mice. These experiments help elucidate the molecular and cellular responses to radiation exposure, including DNA repair mechanisms, oxidative stress responses, and radiation tolerance mechanisms.

Synthetic Biology Approaches: Synthetic biology approaches involve engineering organisms with specific traits or genetic modifications to enhance their radiation resistance or metabolic capabilities. These engineered organisms serve as experimental models to study the effects of radiation on biological systems and develop biotechnological applications for radiation resistance and remediation.

Spaceflight Experiments: Spaceflight experiments, conducted aboard spacecraft and space stations, expose organisms to the space environment, including cosmic radiation, microgravity, and vacuum conditions. These experiments provide insights into the effects of space radiation on living organisms and the potential for biological adaptation and survival in space.

These earth-based models and analogs provide valuable platforms for studying Radiotrophic organisms, radiation biology, and astrobiology-related research questions in a controlled laboratory setting or natural environments with elevated radiation levels. They contribute to our understanding of the mechanisms, adaptations, and ecological implications of radiation-resistant life forms on Earth and in potential extraterrestrial habitats.

Cautions, Considerations and Biosafety

The discovery of Radiotrophs in deep space and their potential return to Earth (e.g. for radioactive contamination remediation or research) would indeed raise important considerations regarding biosecurity, environmental impact, and potential risks to terrestrial ecosystems and human health. While Radiotrophic organisms may offer valuable insights into astrobiology and the limits of life in extreme environments, caution should be exercised to mitigate any potential risks associated with their introduction to Earth's biosphere.

Containment and Quarantine: Strict containment measures should be implemented to prevent the accidental release of Radiotrophic organisms into the environment. Specialized containment facilities, such as biosafety level (BSL) laboratories, may be required to safely handle and study these organisms under controlled conditions. Quarantine protocols should be established to isolate and monitor any returned samples to prevent unintended exposure or contamination.

Risk Assessment: Thorough risk assessments should be conducted to evaluate the potential environmental, ecological, and health risks associated with the introduction of Radiotrophic organisms to terrestrial ecosystems. This assessment should consider factors such as potential competition with native organisms, ecosystem disruption, and the spread of novel traits or pathogens.

Regulatory Oversight: International regulatory frameworks, such as the 50 instruments of ratification, accession, approval, and acceptance of the Cartagena Protocol on Biosafety to the Convention on Biological Diversity (CBD), the Biosafety Clearing-House (BCH), and national biosecurity regulations, may govern the import, handling, and release of extraterrestrial organisms on Earth. Compliance with these regulations and guidelines is essential to ensure responsible and ethical research practices and to minimize potential risks.

Ecological Impact Studies: Prior to any intentional introduction or release of Radiotrophic organisms into terrestrial environments, comprehensive ecological impact studies should be conducted to assess the potential consequences on native biodiversity, ecosystem dynamics, and ecosystem services. These studies should evaluate the compatibility of Radiotrophic organisms with existing ecosystems and the potential for ecological disruption.

Public Engagement and Communication: Transparent communication with the public and stakeholder engagement are crucial to address concerns, build public trust, and ensure informed decision-making regarding the handling and study of Radiotrophic organisms. Public dialogue should include discussions on risk management, ethical considerations, and the broader implications of astrobiology research.

Biosecurity Protocols: Stringent biosecurity protocols should be implemented to prevent the unintentional release or dissemination of Radiotrophic organisms from research facilities or containment facilities. These protocols may include physical containment measures, personnel training, and emergency response procedures to minimize the risk of accidental exposure or escape.

Ethical Considerations: Ethical considerations, including the potential impact on indigenous communities, cultural heritage, and the moral implications of manipulating extraterrestrial life forms, should be carefully considered in all aspects of research and decision-making related to Radiotrophic organisms.

Overall, the discovery and study of Radiotrophic organisms from deep space represent a unique opportunity to advance our understanding of astrobiology and the potential for life beyond Earth. However, responsible research practices, stringent biosecurity measures, thorough risk assessments, and transparent communication are essential to mitigate potential risks and ensure the safe and ethical conduct of research involving Radiotrophic organisms.

Conclusion

In this journey through the depths of space and the examination of potential Radiotrophic life, we have delved into the realm of the unknown, where conventional notions of life as we know it are challenged and expanded. The discovery and exploration of Radiotrophs, hypothetical organisms capable of harnessing radiation as their primary energy source, have opened new avenues of inquiry into the nature of life beyond Earth.

Throughout this book, we have reviewed the resilience and adaptability of Radiotrophic organisms, thriving in environments of extreme radiation where traditional life forms would falter. From their origins in the cosmic void to their potential role in shaping the ecosystems of distant worlds, Radiotrophs offer a glimpse into the possibilities of life in the universe.

As we contemplate the implications of Radiotrophic life forms existing in deep space, we are reminded of the vastness and diversity of the cosmos. The discovery of such organisms challenges our understanding of the boundaries of life and invites us to reexamine our preconceptions about the conditions necessary for life to emerge and thrive.

Looking ahead, the exploration of Radiotrophic organisms holds promise for advancing our understanding of astrobiology and the search for extraterrestrial life. By studying the biochemical and evolutionary mechanisms underlying Radiotrophic metabolism, we may uncover new insights into the fundamental principles of life and its potential manifestations across the cosmos.

While purely speculative, this hypothetical scenario illustrates how a life form adapted to survive predominantly on radiation could hypothetically function. Further research and exploration would be needed to validate and expand upon these proposed concepts.

References For Further Reading

"An ammonia trail to exoplanets", by Andreas Jäger, ETH Zurich Department of Physics, November 7, 2023.

"Astrovirology: Viruses at Large in the Universe", Berliner, A. J., Mochizuki, T., and Stedman, K. M., Astrobiology, vol. 18, no. 2, pp. 207–223, 2018. doi:10.1089/ast.2017.1649.

"The Cartagena Protocol on Biosafety". Unit Biosafety (13 November 2019). The Biosafety Clearing-House (BCH).

"Chemistry of the Elements (2nd ed.)", Greenwood, Norman N.; Earnshaw, Alan (1997). Butterworth-Heinemann. ISBN 978-0-08-037941-8.

"Dark Power: Pigment seems to put radiation to good use", Castelvecchi, Davide (May 26, 2007). Science News. Vol. 171, no. 21. p. 325. Archived from the original on 2008-04-24.

"Dynamics of the universe and spontaneous symmetry breaking", Kazanas, D. (1980). The Astrophysical Journal. 241: L59–L63. doi:10.1086/183361

"Engineering the Genetic Code – Expanding the Amino Acid Repertoire for the Design of Novel Proteins", Budisa, N. (2005), Wiley-VHC Weinheim, New York, Brisbane, Singapore, Toronto

"Evaluating changes in growth and pigmentation of Cladosporium cladosporioides and Paecilomyces variotii in response to gamma and ultraviolet irradiation", Bland J, Gribble LA, Hamel MC, Wright JB, Moormann G, Bachand M, Wright G, Bachand GD. Sci Rep. 2022 Jul 15;12(1):12142. doi: 10.1038/s41598-022-16063-z. PMID:35840596; PMCID:PMC9287308.

"Growth of the Radiotrophic Fungus Cladosporium sphaerospermum aboard the International Space Station and Effects of Ionizing Radiation", Graham K. Shunk, Xavier R. Gomez, Christoph Kern, Nils J. H. Averesch bioRxiv 2020.07.16.205534; doi: https://doi.org/10.1101/2020.07.16.205534

"How do bacterial cells ensure that metalloproteins get the correct metal?", Waldron, K., Robinson, N., Nat Review Microbiology 7, 25–35 (2009). https://doi.org/10.1038/nrmicro2057

"The inflationary universe: the quest for a new theory of cosmic origins", Guth, Alan H. (1997). Basic Books. pp. 186–. ISBN 978-0201328400. May 1, 2011.

"Influx of nitrogen-rich material from the outer Solar System indicated by iron nitride in Ryugu samples", Matsumoto, T., Noguchi, T., Miyake, A. et al, Nat Astron (2023). https://doi.org/10.1038/s41550-023-02137-z

"Ionizing radiation changes the electronic properties of melanin and enhances the growth of melanized fungi", Dadachova E, Bryan RA, Huang X, Moadel T,

Schweitzer AD, Aisen P, Nosanchuk JD, Casadevall A., PLoS One. 2007 May 23;2(5):e457. doi: 10.1371/journal.pone.0000457. PMID: 17520016; PMCID: PMC1866175.

"Ionizing radiation: how fungi cope, adapt, and exploit with the help of melanin", Dadachova E, Casadevall A. Curr Opin Microbiol. 2008 Dec;11(6):525-31. doi: 10.1016/j.mib.2008.09.013. Epub 2008 Oct 24. PMID: 18848901; PMCID: PMC2677413.

"Lead(II) Binding in Natural and Artificial Proteins. Lead: Its Effects on Environment and Health", Cangelosi, Virginia, Ruck thong, Leela and Pecoraro, Vincent L., edited by Astrid Sigel, Helmut Sigel and Roland K.O. Sigel, Berlin, Boston: De Gruyter, 2017, pp. 271-318. https://doi.org/10.1515/9783110434330-010

"Metallomics and the Cell. Metal Ions in Life Sciences", *vol 12.*, Banci, L., Bertini, I. (2013 In: Banci, L. (eds) Springer, Dordrecht. https://doi.org/10.1007/978-94-007-5561-1_1

"Melanin pigments of fungi under extreme environmental conditions (Review)", Gessler, N.N., Egorova, A.S. & Belozerskaya, T.A., Appl Biochemistry Microbiology 50, 105–113 (2014). https://doi.org/10.1134/S0003683814020094

"NASA's Webb Identifies Methane In an Exoplanet's Atmosphere – James Webb Space Telescope", blogs.nasa.gov, 2023-11-22.

"Nitrogen as a Tracer of Giant Planet Formation. I. A Universal Deep Adiabatic Profile and Semianalytical Predictions of Disequilibrium Ammonia Abundances in Warm Exoplanetary Atmospheres", Kazumasa Ohno, and Jonathan J. Fortney, March 22, 2023, American Astronomical Society, The Astrophysical Journal, Volume 946, Number 1

"Prevalence of Earth-size planets orbiting Sun-like stars", Petigura, Eric A.; Howard, Andrew W.; Marcy, Geoffrey W. (October 31, 2013). Proceedings of the National Academy of Sciences of the United States of America. 110 (48): 19273–19278.

"The Quest for Extraterrestrial Life: What About the Viruses?" Griffin, Dale Warren (14 August 2013), Astrobiology. 13 (8): 774–783. Bibcode:2013 AsBio. 13..774G., doi:10.1089/ast.2012.0959. PMID 23944293.

"The spectrum of Jupiter's Great Red Spot: The case for ammonium hydrosulfide (NH_4SH)", Mark J. Loeffler, Reggie L. Hudson, Nancy J. Chanover, Amy A. Simon, December 16, 2015, NASA Goddard Space Flight Center, New Mexico State University Department of Astronomy, USA

"Stability of non-proteinogenic amino acids to UV and gamma irradiation", Rowe, L., International Journal of Astrobiology, vol. 18, no. 5, pp. 426–435, 2019. doi:10.1017/S1473550418000381.

"A stochastic process approach of the drake equation parameters", Glade, N.; Ballet, P.; Bastien, O. (2012). International Journal of Astrobiology. 11 (2): 103–108. arXiv:1112.1506. Bibcode:2012 IJAsB..11..103G. doi:10.1017/S1473550411000413. S2CID 119250730.

"Very High Background Radiation Areas of Ramsar, Iran: Preliminary Biological Studies". Ghiassi-nejad, M.; Mortazavi, S. M. J.; Cameron, J. R.; Niroomand-rad, A.; Karam, P. A.. Health Physics 82(1):p 87-93, January 2002.

"Webb Discovers Methane, Carbon Dioxide in Atmosphere of K2-18b", nasa.gov, September 11, 2023, NASA Webb Telescope Team.

1 million	=	10^6	or	1,000,000
1 billion	=	10^9	or	1,000,000,000
1 trillion	=	10^{12}	or	1,000,000,000,000

1 light-year	=	9.46 trillion km or 5.88 trillion miles
1 parsec	=	3.26 light years

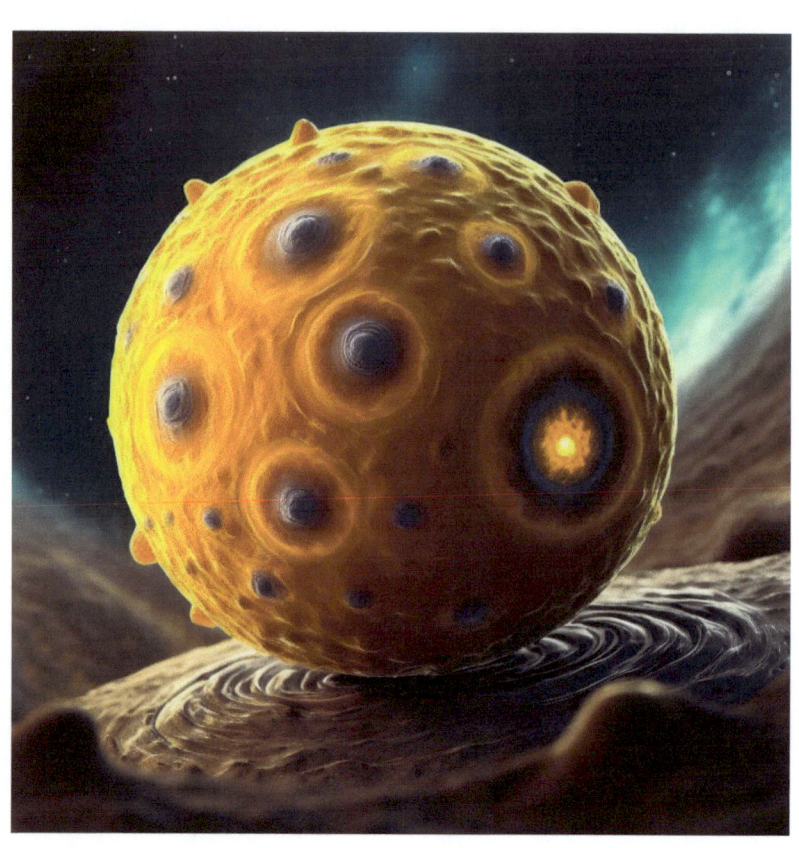

www.ingramcontent.com/pod-product-compliance
Lightning Source LLC
Chambersburg PA
CBHW042038230526
45474CB00005B/4